Horse Drawn Equipment
&
Early Tools

Haywood County North Carolina-
Great Smoky Mountains

Best Wishes
Louise K. Nelson

Louise K. Nelson

Publisher: Louise K. Nelson

Cover Design: Walsworth Publishing Company, Inc.

Interior Design: Susan Allen

Printed in The United States of America

ISBN 0-9671476-0-3 Library of Congress Number 99-093129

The book may be purchased from dealers or by calling or writing the author.

Ordering information page 73

Other books available - information page 73

Contents

Horse Drawn Equipment

and

Early Tools

Haywood County, N. C. & Great Smoky Mountains

Dedication:

In honor of the ones supplying the photos and in memories of their families who once owned a large portion of the equipment and tools.

To John Francis and the Francis families before him.
To Fran and Terry Rogers and to the Nolands, Rogers' before them.

Thanks to:
Gordon Anderson for identifying the photos
Raymond O'Neil for his story
King Family for using a picture of their old home
Pisgah View Ranch for use of their home photo
Randy Nelson for drawing of tenant home on McCrary Farm, Big Branch, N.C.
Wilson Medford for the use of his father, W. Clark Medford's early history books.
Magnolia Antique Mall for use of their pictures.

Some of the pictures we haven't been able to identify. With this collection I hope you will have a good history of the early tools and horse drawn equipment. Plus information of the early settlers of Haywood County, North Carolina and how the settlers lived in these years in the Great Smoky Mountains, also.

Introduction

Horse Drawn Equipment & Tools came about as a series of books that I am writing about this era of life.

After starting a few years ago with my cousin, Jack Caldwell, in search of this old equipment and implements I found it very interesting. It too, will be a history of the past to hand down to our grandchildren and great- grandchildren.

In writing I thought it would be best to explain how the pioneers arrived and how they survived with what few tools, oxen or horse that they brought with them when they started in Haywood County, North Carolina.

I used Haywood County since this is where I have always lived and it is easy to get the necessary information from folks I know.

All Western North Carolina to Great Smoky Mountains started the same, lived the same and used the same type of equipment and tools.

To the best of my knowledge and with the help of other people we have tried to be as accurate as we possibly can. However, if some of the information is incorrect then I'm sorry.

I feel the information collected will be enough to cover most of the tools and equipment in the book.

The First Settlers

The first settlers coming to Buncombe County, North Carolina, (which became Haywood County North Carolina, 1810) and to Mount Prospect (which became Waynesville, North Carolina) came with a few tools, pots, pans, vegetable seeds for planting, flour, meal, milk cow, oxen, horses, handmade sled or cart, Bible, gun, gun powder, an axe, dog, clothes, shoes and the bare necessities.

They came from Ireland, Scotland, England, Whales, Germany and other countries.

Indians

These pioneers came into a wilderness with many wild animals. The Indians were living in this county in 1540 when Hernando Desoto came through looking for gold. How long they had lived here no one knew.

A group of two thousand soldiers with General Griffith Rutherford were sent here by the British Agents in 1776 to move the Indians to an area all by themselves to make room for the white settlers. This area became Cherokee, North Carolina.

Move To Haywood County

When the Revolutionary soldiers moved to Haywood County they were given land. By marking, clearing off the land, building a shanty log home and entering it into the Entry Book of Waynesville, it became theirs.

When the town of Waynesville was in need of a jail and a court house a tax was added to the land. Many settlers lost acres because of lack of money to pay the tax.

Court had been held in private homes in Waynesville and in the Buncombe County Court House. All important matters had to be taken to the Buncombe County Court House.

Waynesville had slaves, bound boys and girls, whippings and hangings for the wrong doings.

Waynesville had to establish law and order and they must do it correctly.

Waynesville soon had lawyers to help with the legal matters.

Log Homes, Equipment, Barns, Fences and Tools

Pioneers worked daily cutting and removing trees to make room for a log cabin and to have a spot to farm. Their tools weren't good for cutting trees. There were stumps and rough looking parts of the trees standing for years.

They used the wood to build cabins, barns, fences, equipment, hand held tools and to burn in the mud and rock fire place.

The fences were necessary to ward off some of the wild animals from their livestock.

Settlers old homes possibly looked like this.

Log Buildings of The Past

This nice old log home was built 1790. It has been well taken care of. Small dwelling in the back. Located at Pisgah View Ranch, Candler, North Carolina. Courtesy of the Pisgah View Ranch. Old wagon with wood in the yard.

This old log home was occupied by my Grandparents, the Kinslands, when they first moved to Crabtree, in Haywood County in the early 1800's. This shows how wood and rocks were used.

This old log home was the tenant home of Wiley and Dolly Caldwell, grandparents of author, Louise K. Nelson. The smallest log home was over 100 years old when Louise lived with her grandparents here. The floor was partly dirt floor near the fireplace. Artist drawing by Randy Nelson from the memory of Louise as to how the old home looked.

This old log barn was built from the many logs of the past and is sitting on rocks for a foundation. The door from planks was added later.

This old log cabin is still standing 1998, owned by the King family from Hominy, N.C. The lean to of the house was later built from planks which became the kitchen. The cabin has a rock foundation and chink mud held the logs together.

Wild Animals

In the wilderness there were many wild animals such as: coyotes, fox, bears, wildcats, panthers, wild boar, deers, turkeys, wolves, and others. There were many small animals as: squirrels, rabbits, ground hogs, opossums, skunks and many more. Plus all this there were many snakes. A lot of the animals were dangerous. There were also many animals that were used for meat. In fact, the settlers had plenty of meat.

For many years the settlers had to combat with the wild animals.

Hunting Dogs

Every household had dogs. They were very useful. They were trained to herd cattle and sheep. They were trained to chase away wild animals. They were also used in hunting animals for meat and to hunt animals used for hides to sell.

Animal Hides

Soon the pioneers learned to dry animal hides to make clothes, bridles and saddles. Also to make shoes.

Tom Gaddy from Ratcliff Cove had made old-fashion brogan shoes for years. Then his son Bill Gaddy and Bob Underwood made shoes before Bill became one of Waynesville's first barbers.

Spinning Wheels

The women were busy carding the sheep wool, making it into thread on the spinning wheel to make clothes. They had learned a lot as they went along.

Streams, Branches and Rivers

The county and town were well provided with good clean, fresh, rippling waters from the many spring heads coming from the mountains and being headed up to make springs and into a log spring house with a trough to cool the milk, keep the butter and foods that needed to be kept cold. The streams and the branches were full of mountain trout which made a good source of meat. The branches supplied water for the livestock and flowed on into rivers.

Pioneers Were Progressing

The pioneers were learning more and more how to grow corn, wheat, rye, potatoes and vegetables. They had to save seeds from their crops and have the woods cleared as the planting season began. Many, many of their tools were made from wood.

They were taking advantage of the wood to make barns, sheds, spring houses, smoke houses for storing meats, corn cribs, chicken houses, cellars and toilets. They knew if they kept the hogs pinned up they could fatten them faster, so they built hog pens to keep them in. All this took special tools but they had learned to make them.

Eventually they built log homes with an upstairs and more rooms. They even had windows. They used their special tool to make the boards for the roofs.

After many years the town of Waynesville, N. C. had tanneries. The hides were used for many things, some being saddles, bridles and shoes. Grist mills had started throughout the county to grind corn into cornmeal and wheat into flour. Until now they had used a handmade grinder to grind the corn into meal to use for making bread.

They had used corn shucks and straw to make bedticks (mattresses, now days). Knitting mills had opened up and they could buy some material to make clothes. The farmers were growing apples, making molasses from cane and they had bee hives to hold the honey bees for making honey. The farmers were also growing tobacco and they were able to sell this at the tobacco warehouse now in Waynesville. They had livery stables and blacksmith shops. The grannies (mid-wives) were delivering babies until the new settlement had doctors. One-room schools had started to educate the children and there were churches. Little by little they were able to purchase some medicine. Until now they had depended on herbs, spices and roots for their medicines. Saw mills had begun and the oxen and horses pulled the logs into it to make lumber.

Buggies, wagons used in riding and carriages were giving way to the cars that hit some of the area by 1925.

The cattle farmers had drovers to drive their cattle to market. After many years the cattle market was in Clyde, N. C. and the cattle could be shipped out on the new trains. The drovers were still traveling through the woods and the logging roads and across the Pigeon River to reach Clyde, North Carolina.

A few country stores were in effect. They transported a lot of the necessary items by wagon to the stores and in return they carried items to the market to exchange for these items.

After many years a wood stove for cooking had replaced the fire place and this was a good means of using their wood.

All this was great but the farm horse drawn equipment and hand made tools had it's place in the lives of the early pioneers.

Imagine they came to this new settlement with the bare necessities. I can imagine how they guarded their guns, axes and knives.

The plows, harrows, sleds and wagons were made of wood. Some of the early items used were wooden nails, then nails were made at the blacksmith shop. Some of the early tools that were mentioned were: augers, broad axe, pole axe, single and double bladed axe, hatchet chopping axe, mallet, maul, froe, board splitter, foot adz, hammer, hoes, one-eyed chopping hoe, mattock, handmade tobacco and corn cutters, sycthe, crowbar, plow stocks, oxen yokes, horse shoes, shavers, chisels, go-devils, square, ruler, wooden nails and more.

There were drawing knives and shaving horses. They used a rock called a whet to sharpen their knives. Some of the equipment and tools mentioned were being made at the blacksmith's shops.

Cross-cut saws became popular as time went on. This was a big improvement for the early pioneers.

From locust wood they made fence post and rail fences, wagon pillars, and axles. Walnut wood was a hardwood with good color and grain. It was used in building furniture, wheel spokes and roofs. Pine was a softer wood. It was used in making doors and door frames. The pine knot with the resin was used to build fires. Maple wood was used to make furniture, also fiddles and guitars. Cherry wood was used in furniture and etc. Hickory was a hardwood used in oxen yokes, wheel hubs, wagon tongues and axles. Chestnut was used to make handles for tools and coffins. Poplar wood was also used to make coffins, axles, weatherboard, furniture, rafters, joists and logs for building. Black gum was used to make bee gums. Wagons and sleds were made from a hardwood.

The womenfolks had to have many handmade items for the home. Some of the items were: spinning wheels, dough boards, bowls, spoons, plates, churns, buttermolds, barrels for pickling vegetables and storing meal and flour, wash tubs, wash pans, dish pans, kitchen utensils, furniture as: beds, dressers, tables, safes for dishes, benches and chairs. The chairs were bottomed with the bark that came from white oak trees. The baskets were made from the white oak bark.

Tomato and tobacco sticks were made from wood.

When killing hogs the scaffold was made from logs as was the pole used in making molasses. The mules pulled the pole around in circles to squeeze the juice from cane in making molasses.

Rock was used in many ways as foundations for homes, barns and buildings put together with mud.

Western North Carolina and The Great Smoky Mountains were slow in getting the modern equipment. So many people did not have the necessary money to buy them. The early modern equipment was still pulled by oxen, horses or mules.

In the collection of Horse Drawn Equipment and tools you will see a lot of the early modern equipment.

Estate Settlement Of Farm Stuff-1810
W. Clark Medford's Early History Book of Haywood County

1 bell & collar	$ 1.50
1 broad axe	1.00
1 iron wedge	1.00
cutting knife and pack saddle	2.00
1 bar shear plow	4.00
2 cows with bells on	24.00
1 deer skin	.50
1 side of leather	.87
1 cow and calf	10.50
1 churn	1.12
1 melting ladle	1.12
1 shovel plow	3.00
8 drakes	1.00
21 shoats	36.00

The Francis'

The Francis' arrived in Haywood County and Waynesville, N. C. around 1818.

William Francis, one of the most noted builders came from East Tennessee and settled in Francis Cove, N. C.

William Francis along with his son, Henry, built a number of grist mills in Haywood, Buncombe and Jackson County.

They also farmed, apple orchardmen and even tried their luck with silk worms.

Pinkney Francis prided himself in building a lot of business buildings and homes in Haywood County and Waynesville.

Pinkney was a true craftsman.

The Francis's each had their own trade; dairy men, carpenters, farmers, orchardmen and Pinkney spent a large part of his life working at the Waynesville Tannery.

Robert (Bob) Francis, father of John Francis, collected many pieces of old horse drawn equipment from early 1900s. Most of it coming from the mountains of Haywood County.

He had a true feeling for the folks of Haywood County and for the old horse drawn equipment.

Bob not only farmed, he raised pigs, cattle, and turkeys. The horses and mules had a special place in his life.

In the early 1900s Pinkney Francis built the Wood Lathe, listed as (1). He also built the Loom listed as (2) in the early 1900s.

Horse Drawn Equipment and Hand Held Tools by John Francis

(1) Wood Lathe-built around 1900 by Jeremiah Pinkney Francis. Lathe is powered by the right foot and has 3 different speeds. One would work.

(2) Loom - built around 1900 by Jeremiah Pinkney Francis.

Covered Doctor's Buggy

Covered Wagon - A wagon used often by Robert Henry Francis for wagon trains and enjoyment.

Uncovered Doctor's Buggy

3 Seated Surry

2 Seater Surry - The front seat backs would roll from front to back allowing the passenger to either face the road or passengers.

Hay Rake - After dropping the rake to the ground and the rake got full of hay, you would trip the rake leaving the hay behind and starting again. The hay left behind would be picked up by sled or wagon and taken to the barn.

Hay Raker and Mower

Oliver Mowing Machine - Not a very popular mowing machine in this part of the county.

Grist Mill - used to make cornmeal. Drop corn in top and stones would grind the corn into fine cornmeal. The belt in the picture would be connected to a hit and miss engine.

Planter

Horse Drawn Grain Drill - Once owned by the Boone family of Waynesville, used to put small grains in the ground as grass seeds and clover.

Threshing Machine - Used to separate wheat or rye from the stem.

Riding Turning plow

Riding Cultivator - used to cultivate rows of crops for weeds and grass.

Foreground - Apple Cider Mill. Background - Garden Sprayer-pumped by man power, by pumping on the lever, spray would come out the hose.

Orchard Sprayer - Used by Barber's orchard during the 1920s through 1940s. Sprayer was drawn by mules who were very accustomed to the hit and miss engine.

Apple Crate Sled - a replica of the one used by Henry Francis in the late 1930s-1940s, to bring apples off the mountain at the Francis Fruit Farm. Most sleds were made right there on the farm.

Horse Drawn Oliver Superior Manure Spreader

Stationary Molasses Mill - The horse or mule was hooked to the beam and caused to walk in a circle around the mill. The mill would squeeze the syrup from the cane stalk and drain into a covered bucket.

Turning plows - Various turning plows - each was used for different purposes.

Foreground - one row cultivator
Left background - one row corn planter

Right background - turning plow
center background - lay off plow

Cultivators and Turning Plow

Grabs - pictured are trail & butt grabs. For wood logging
Trail grabs - used if you hook more than one log together. 2nd. log trails the first. Butt grabs - put in the butt of the log and hooked to the horses.

Cross-cut Saws - Hanging are spreads used in logging, stick is missing. A spread is connected to the log, bottom of page.

Horse & Mule Hames - the ones to the far right and left are orchard hames. Notice they have no knobs to knock the apples from the trees. Orchard hames are also called beagle hames.

Various size horse or mule collars.

Double trees - Used when two horses are working together.

Early 1900s Horse Saddles.

Drag Pan-center against wall-Mules or oxen were hooked to the pan while a person scooped up dirt to move to another location. Used on main street and in digging out basements when the stores in Waynesville were being built 1910-1920
right-Cream Separator 1950s

Far left- rear axle for a wagon. Center-various wagon wheels.
Hanging- various ox yokes

A replica of the Early Stage Coach

Fertilizer cart (the cart is upside down in this picture) - Used in the Waynesville Fertilizer plant during 1920s, 30s, and 40s to move fertilize from one location to another. The fertilizer plant had 20 or more of these carts.

Farm Wagons - Used to transport corn in and out of the fields, haul rocks and whatever else was needed moving on the farm. Note: All mountain wagons had brakes on the rear wheels.

McCormick Deering International # 9 high gear mowing machine. This model was by far the most popular model for the mountain terrain, because of its ease of pulling and fast cut.

Thornburg Wagon - very popular and appreciated.

Collection of Horse Drawn Equipment and Tools - With the help of Jack Caldwell

Handmade Sled

Wooden Harrow and rail fence

Discovered this old planter that had been uncovered from a pile of dirt - Macon, Georgia Flea Market

Corn Crusher - compliments of Jack Caldwell

Old planter found stored in a barn for many years

Hay Spreader - Kelly's Farm
Big Branch - Haywood County

Mowing Blades, pitch Fork from Noah's
House - New Hampshire

Collection of old tools - Sam Cable's

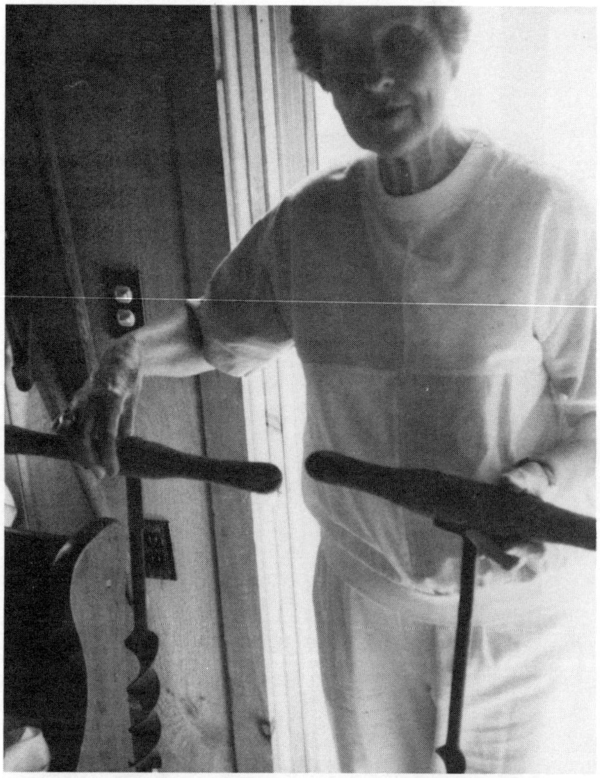

From Noah's House - New Hampshire

Old saws - from Noah's House - New Hampshire

Cross Cut Saw and Mowing Blade - Doyle Brown

One of the many horse drawn plows - Jack Caldwell

This collection of old saws and hand held tools - compliments of Magnolia Mall - Waynesville, N. C.

Top - scraper
Middle - lumber scale
Bottom - protractor square

From Jack Caldwell

This group of hand held tools and horse drawn farm equipment is from Fletcher Farm Fair 1998. They are described by Gordon Anderson.

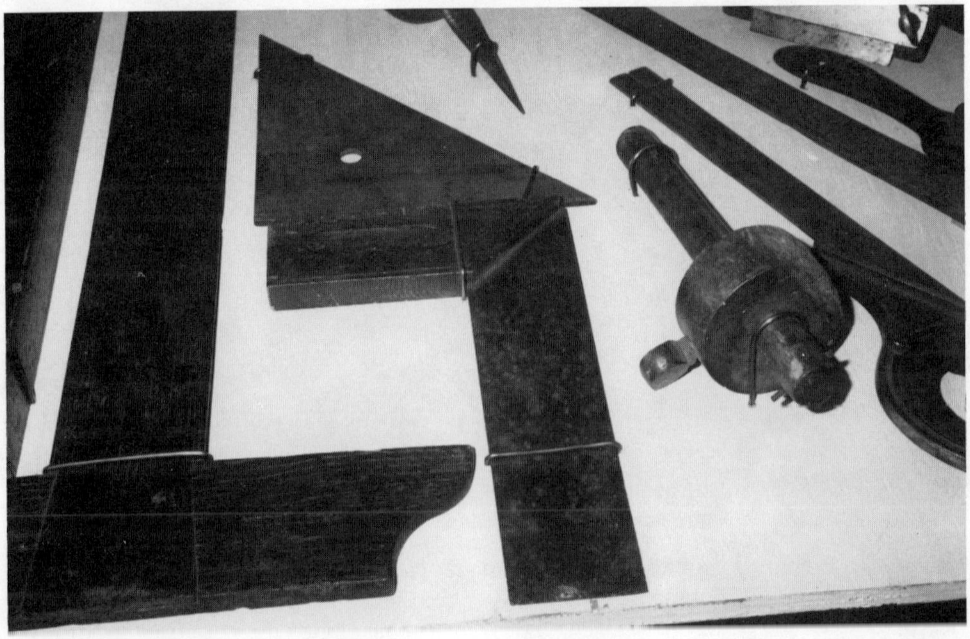

Left - wood square, top center - draftmans square, try square, right - adjustable scriber

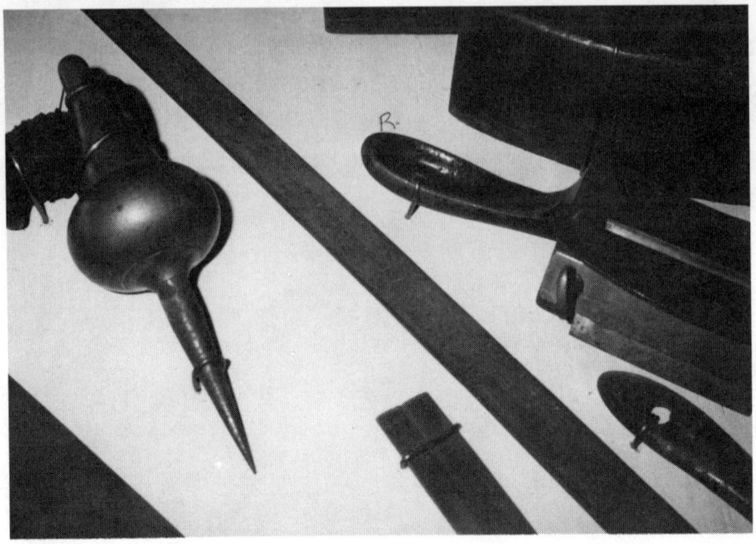

- Plum Bob
- Lumber rule
- Scraper for smoothing wood

top - wood block plane, bottom - scraper for smoothing wood.

bottom left - molding plane, center - ruler with angle blade, lumber rule, right - level

top left - axe heads, draw knife, chisels, center - claw hatchet, right - broad axe

auger bits

Left to right, cow bells, charcoal fired iron, hand seed cranker, right-one-leg milk stool

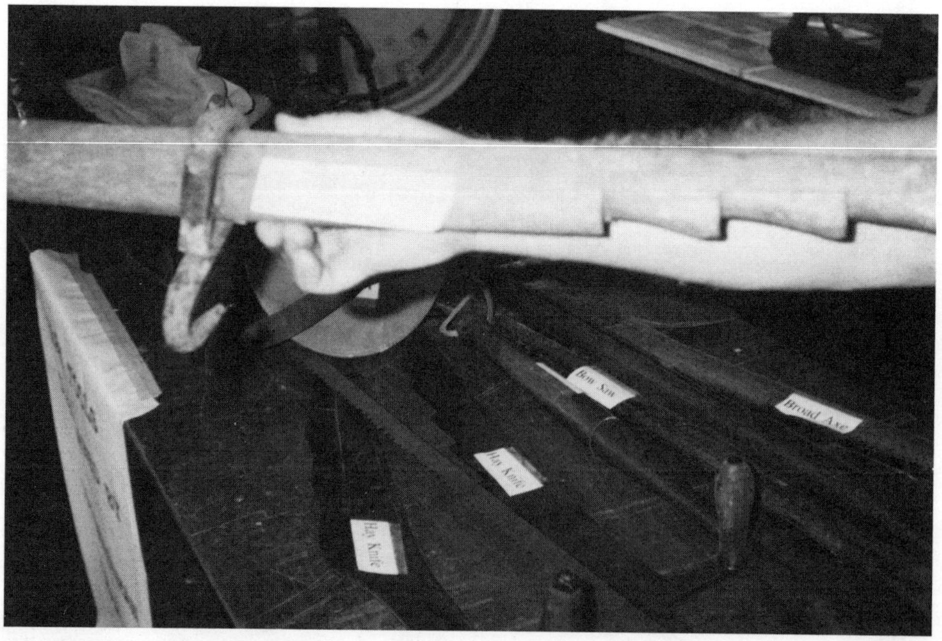

Top - used in hanging animals to dress after killing them. Bottom - hay knives, bow saw and broad axe.

Hay Knives

One-man Cross Cut Saw

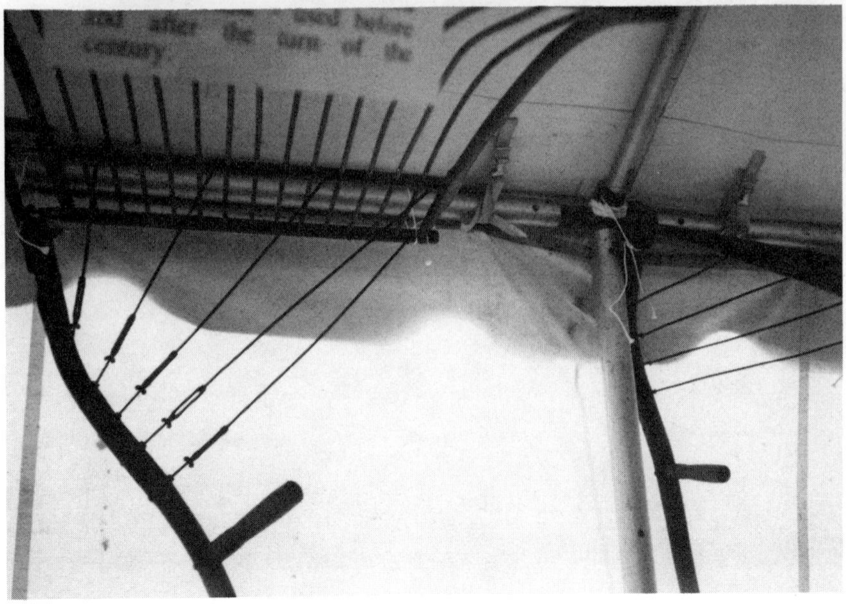

Wheat Cradle
Reaping Cradle

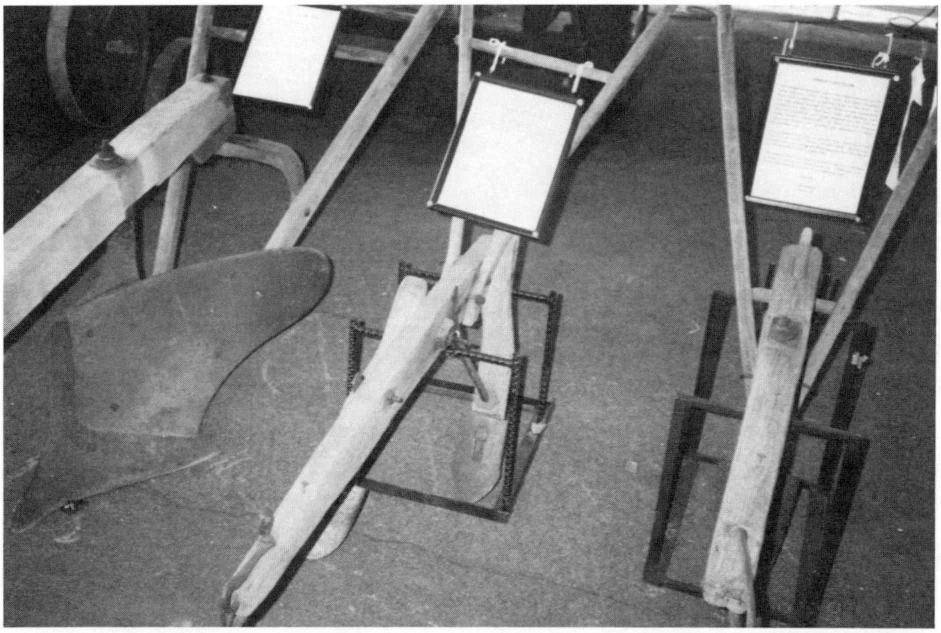

Plows - Fletcher Mountain Fair - 1998 - Early steel plow, double shovel plow, single shovel plow

Old Reaper

Riding Cultivator

Horse Powered Hay Baler - the horse was hooked to the beam in the foreground and caused to walk in a circle. The hay baler was stationary. The hay was brought in from the fields and manually put into the machine.

Robert (Bob) Francis and his father, Robert (Henry) Francis 1946

Bob Francis & Tim Whitner - July 1981, horses Jerry & Kate, Fannie & Belle
Bob loved to travel with the wagon trains.

Jack McCracken, Susan Francis' grandfather.

Since Bob loved his farm and his horses, the family decided to drive the body to be buried from the funeral home to the cemetery. His son, Bill, was driving. August, 1991

Oxen & Men logging wood at Hazel Creek, N. C. Year unknown

Courtesy Of Tim Kelly
Plant Setter

Horse Drawn Farm Equipment from the Tim Kelly Farm - Big Branch

Planter

Saws

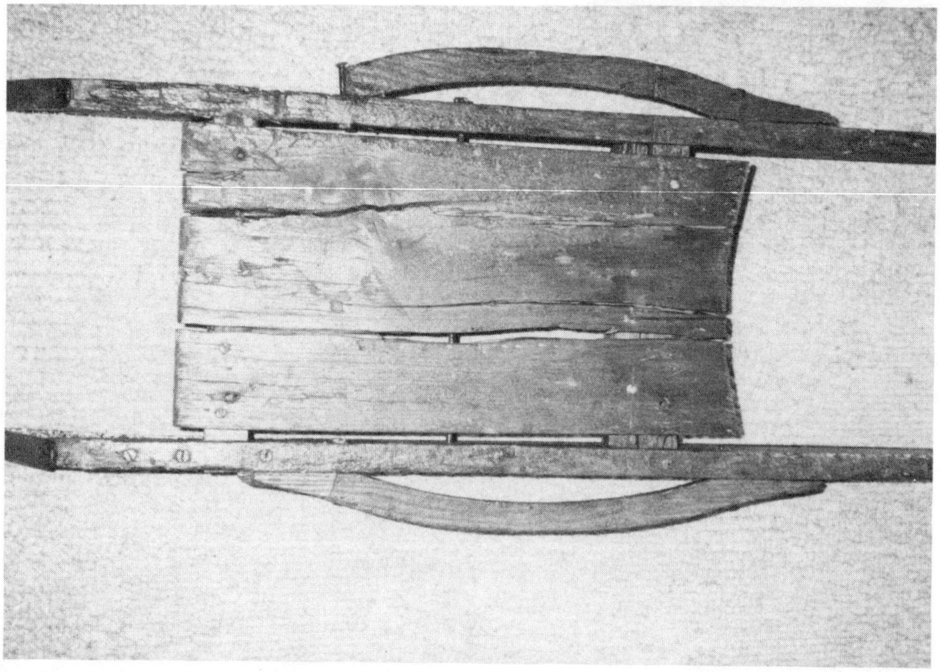

Child's Handmade Snow Sled

Courtesy Of The Waynesville Mountaineer - Centennial Edition. When the farmers used the oxen to pull the wagon of cut hay to the barn or to make a hay stack.

This picture of haystack at Fines Creek, N. C. 1933 - Gus & Joe Ledford, Courtesy of Kathy Swanger.

Tools - Courtesy of Magnolia Mall - Waynesville, N. C.

Traveler's Measuring Tool

Froe & Maul

Rogers' & Nolands

Fran & Terry Rogers' hand made tools and equipment were handed down by Terry's grandfather and father.

Terry's grandfather, Tommy Noland, came to Upper Crabtree from Sandy Mush N. C. when he was 8 years old.

Terry's father, Cassius Rogers, son of Robert and Bly Rogers all were natives of Haywood County and their home was at upper Crabtree.

They were both every day farmers growing what was necessary to feed their families. But, they were also cattle and tobacco farmers. They grew hay for the cattle's winter use and corn to fill the silage for the livestock's winter use.

They went through farming days when they used a lot of the horse drawn farm equipment, and in the latter years they graduated to the modern equipment drawn by the horses and eventually the motorized equipment.

Froe for splitting shingles
Notice the old tree stump with each line representing one year.

Pictures of tools and implements owned by
Terry Rogers Upper Crabtree

Wooden
Wheel-barrow

Wooden Rack is used to move dirt & small rocks.
Wooden Rack - Handle is one piece, split part way.

Lay off plow for rows & tobacco ridges

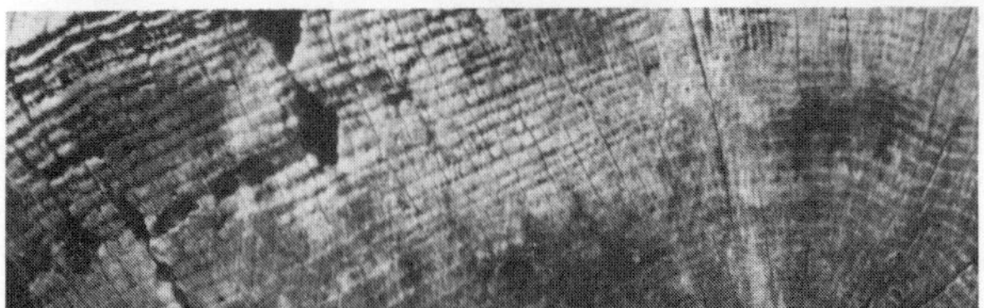

Handmade Corn Shucker

Wooden Drill

Raymond O'Neil and His horse, Buck Shot

Raymond is retired from his job of many years and this gives him time to work around his home at Cruso, North Carolina. He has made and collected many things of the past and enjoys just putting Buck Shot, the sled and plow together in making a garden.

He and his wife enjoy their modern home but they love sitting on the porch looking in the pasture at the steers and watching Buck Shot. The beautiful mountains are a part of their view as they live these years in enjoyment.

This handmade sled was made by Raymond O' Neil many years ago.

Single tree to connect Buck Shot and the sled for working.

Other Important Pictures Of The Past.
Hand Powered, Horse Drawn and Motorized

In 1925 this beautiful car, Ford Sedan, came into Haywood County. What a change! The buggies, carriages and the covered wagons would fade away.

Then there was this hit and miss engine around 1900-1920 power of 1 hp.

Anvil used in the early years when the railroad came to the mountains.

Dewey Francis' Grist Mill - Francis Cove, Waynesville, N. C.

Rotary Grass Mower around 1930

Molasses Heat Evaporator,
Molasses Skimmer

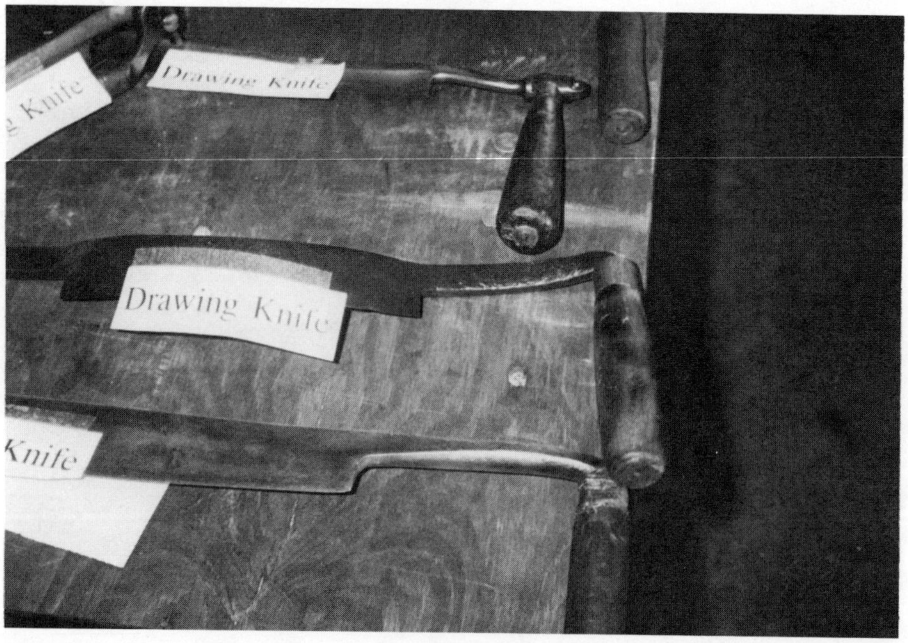

Drawing Knives

One of the first Motorized Hay Bailers - Late 1940s

Corn Binder - Power take off driven, late 1950s

Chicken Coops - used by Francis Farm Mill during 50s & 60s. Used to carry chickens to the market. About this time the hens were not producing as many eggs and they were being sold.

Egg Baskets - Used in the 50s, 60s & 70s. The Francis Farm Mill had about 70,000 chickens and gathered hatching eggs for Frank Perdue. These are collector items from the past.

Silage Chopper - this is a Blizzard Model. The corn was brought from the field and fed into the chopper where it would chop and put the silage to the top of the silo. It was powered by a tractor or a Model T. The belt would run from the tractor's external fly wheel or from the Model T's wheel that had been jacked up.

Courtesy of Mark Palmer

Change In The Barns

Side view of barn - storage in top, possible cow stall below

Old Barn & Silo
A change as years went on.
Pictures courtesy of
Ann Kinsland

As the years went on tobacco became a good source of the farmers income.

In the early years tobacco was not sprayed for worms. Everyone had the job of planting the seeds in the early spring, setting out the plants by hand. Fall of the year when the tobacco was in case it was time to class it and get it ready for the market.

Horses Are Pets Or Pleasure Horses Now-days

Pictures - Courtesy of
Bob Allen

Corn was a major crop. Cornshocks no longer exist.

Fines Creek Corn Shucking

Picture of corn shucking approximately 1915, Fines Creek, N. C. The corn shuckings were also an entertainment time. From left to right, two young boys in back unknown. Left to Right: Joseph Kirkpatrick, Ben Kirkpatrick, Cauley Rogers, France Rogers, Shook Ferguson, Melvin Rogers (father of Cauley Rogers), Joe Kirkpatrick (with fiddle), Joe Kirkpatrick's grandfather, Bob Justice (my relative) with banjo.
Picture donated by Joe Kirkpatrick.

From Horses To Automobile

Photo courtesy of Mountaineer Enterprise.

About The Author

Louise K. Nelson began writing articles for the local paper, Mountaineer Enterprise 1985. After much encouragement from the readers, she decided to put the writings together for her first book, "Country Folklore". This book was published January, 1997 and has been a great success. There will be a series of six books from the pioneer days through 1930s. "Country Folklore" tells how the poor tenant farmers lived 1920s-1930s. November 1998 the second book, "The Aroma & Memories Of Grandma's & Mama's Kitchen" was published. This book too, is a success. This book carries the old cooking methods and recipes for this era of life plus memories of the past.

This book "Horse Drawn Equipment & Early Tools" as it was in Haywood County on into The Great Smoky Mountains from the pioneer days of handmade tools and equipment drawn by the horses, oxen and mules to the more modern equipment still drawn by the horses.

Each book is a history of these days and should be handed down to our grandchildren and great- grandchildren. A part of history that needs to be preserved.

Check ordering page 73 for ordering of each book.

Ordering Information

To Order:

(A). Country Folklore @$12.95 per book plus 6% tax (N. C. residents only) plus shipping ($3 for first book; $1 for each additional book).

(B). The Aroma Of Grandma's & Mama's Kitchen @$12.95 per book plus 6% tax (N. C. residents only) plus shipping ($3 for first book; $1 for each additional book).

(C). Horse Drawn Equipment & Early Tools @$10.95 per book plus 6% tax (N. C. residents only) plus shipping ($3 for first book; $1 for each additional book).

When ordering from out of state, deduct NC Tax from each book.

To Order: List how many

(A)______________ (B)________________ (C) ________________

Name ________________________________ date ____________________

Address ____

City______________________State___________Zip_______________

Total books ________
N. C. Tax __________
Shipping ___________
Total _______________

Mail check or money order to:

Louise K. Nelson
Country Folklore
1336 Asheville Rd.
Waynesville, N. C. 28786

phone 828-456-3760

Book Dealers call or write for prices.